A COLOUR G
TO CLOUI

BY

RICHARD SCORI

Professor of Theoretical Mechan
Imperial College, London

AND THE LATE

HARRY WEXLE

Formerly Director of Researc
United States Weather Bureau
Washington

PERGAMON PRES

OXFORD · LONDON · PARIS · F

THE MACMILLAN COM

NEW YORK

Sun pillar and arc of contact at sunset, prod
in hexagonal ice prisms with

PERGAMON PRESS LTD.
Headington Hill Hall, Oxford
4 & 5 Fitzroy Square, London W. 1

THE MACMILLAN COMPANY
60 Fifth Avenue, New York 11, New York

COLLIER-MACMILLAN CANADA, LTD.
132 Water Street South, Galt, Ontario, Canada

GAUTHIER-VILLARS ED.
55 Quai des Grands-Augustins, Paris 6

PERGAMON PRESS G.m.b.H.
Kaiserstrasse 75, Frankfurt am Main

Library of Congress Card No. 63-19254

By R. S. Scorer and H. Wexler
A COLOUR ENCYCLOPAEDIA OF CLOUDS
CLOUD STUDIES IN COLOUR

with A. K. Showalter
A COLOUR PANORAMA OF CLOUDS (1964)

by R. S. Scorer
NATURAL AERODYNAMICS

Printed in Hungary
Kossuth Printing House, Budapest

187

A COLOUR GUIDE TO CLOUDS

RICHARD SCORER and HARRY WEXLER

The Commonwealth and International Library of Science, Technology, Engineering and Liberal Studies

FIG.

FIG.

Invers

THE numb
erately kep
the reader
to get to k
national La
common E
recent year
Cumulus
cirrus a fibr
tion, no fib
in a layer,
or *-nimbus* n
lus, and *pil*
from the gr
Mamma are
These Latin
example, *cu*
The preci
streets, *therm*
and the pic

Note: In the te
numbers.

THE
thei

The

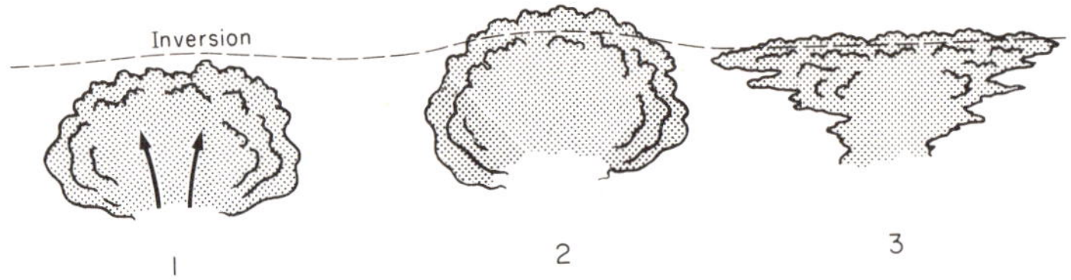

FIG. 5 If a thermal is not buoyant enough to penetrate through an inversion it becomes flattened out at it, as shown in the three stages here. *(See* 3, 7, 8, 14)

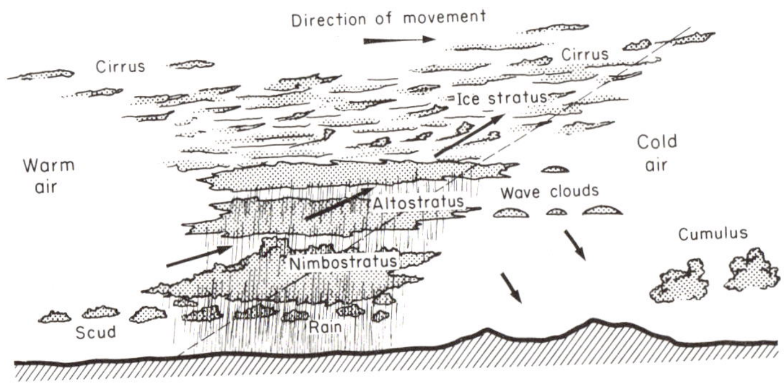

FIG. 6 This cross section of a warm front shows the various layers of cloud which frequently occur. But it must be emphasized that warm fronts show a very great variety of cloud systems. This diagram represents about 300–500 miles in width and about 6 miles in height. *(See* 19, 20, 21, 22)

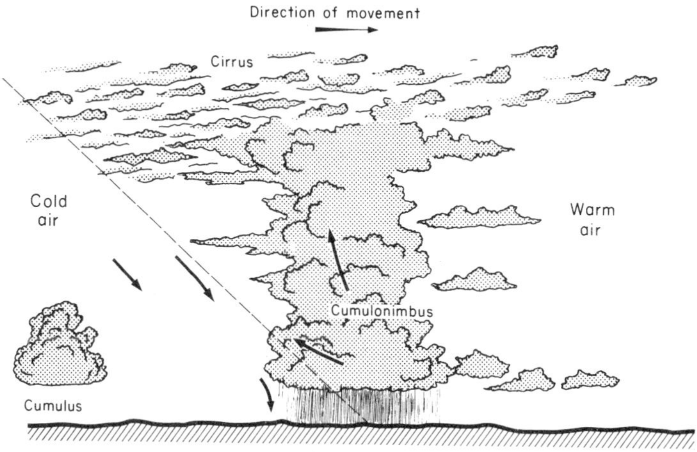

FIG. 7 Cold fronts usually have clouds which are much more like cumulonimbus, as shown here. As in the case of warm fronts, cold fronts can show a very great variety of cloud systems. *(See* 24, 25)

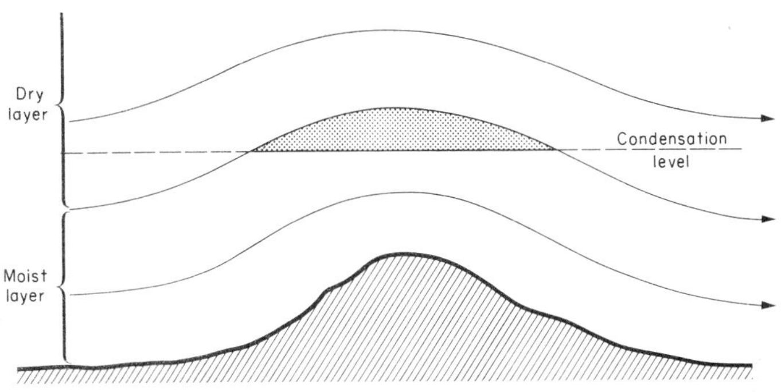

FIG. 8 If a layer of air is carried above its condensation level in passing over a hill an arched cloud is formed, particularly if the air above is much drier and no cloud is formed in it. *(See* 28)

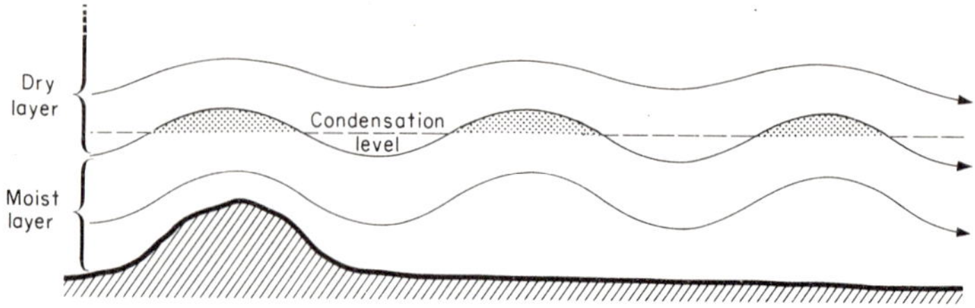

FIG. 9 After passing over a hill the air may continue to oscillate up and down and similar clouds are then produced parallel to the hill, at fairly regularly spaced intervals. These are called lee waves. *(See* 31)

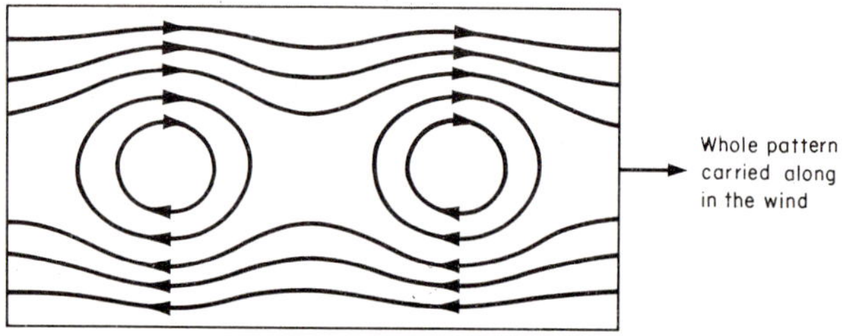

FIG. 10 Sometimes a rolling type of motion occurs in between two layers of air moving at different speeds and the clouds then become arranged as billows. The precise appearance of the clouds depends where abouts they are in this pattern of motion. *(See* 32, 33)

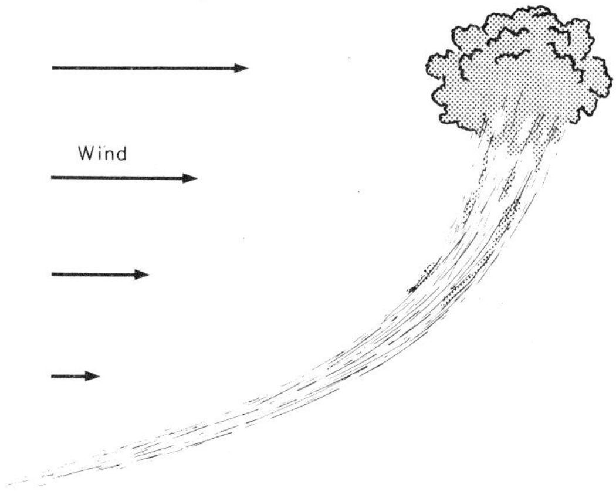

FIG. 11 If a thermal, or for that matter any other kind of cloud has ice particles falling from it and the wind is less at lower levels, those particles which have fallen farthest get left behind and a great sweeping trail is produced. *(See* 35)

LIST OF ILLUSTRATIONS, SHOWING ARRANGEMENT OF SUBJECT MATTER

Forms of Cumulus

The Clouds of a Low Pressure Area (Depression)

Wave Clouds, Billows

27 Cumulus; wave cloud; cirrus
28 Wave clouds; billows
29 Wave hole
30 Cirrus plume
31 Streets; lee waves
32 Altocumulus billows
33 "Breaking" billows

Ice Clouds

34 Cirrus
35 Cirrus fallstreaks
36 Contrails; artificial thermals

Coloured Arcs

37 Mock sun
38 Circumzenithal arc
39 Glory

Cloud Layers

40 Altocumulus
41 Stratus; altocumulus
42 Radiation fog
43 Sea fog; cumulus; stratocumulus

Wet Grass

44 Wet Spider's web
45 Guttation

Violent Storms

46 Tornado
47 Hurricane

Placid Scene

48 Stratocumulus; cumulus

Forms of Cumulus

1. *Cumulus* is the result of *convection*. The ground is heated by the sunshine in the morning, and masses of warm air, called *thermals*, rise. After they have passed their condensation level they become visible as clouds of water droplets, and each cumulus cloud is composed of one or more thermals. The condensation level is at about 3000 ft above the ground on this summer morning in N. England.

   The thermals mix with the surrounding air, which is drier, and consequently the cloud soon evaporates, and the cumulus will only remain so long as thermals continue to rise into them. The rising parts have sharply outlined tops and are very white; the evaporating parts are ragged and of a duller colour.

2. *Cumulus over land* in the morning shows that the land is being heated by sunshine. Out to sea, no cumulus is produced. In this picture, taken from about 16,000 ft near the coast of Belgium, there are a few distant fragments of cloud over the sea left over from the night, but they are not "boiling up" and rapidly changing their form like the land cumulus.

3. *Cumulus over the sea* is common in autumn and winter when the sea is often warmer than the land. This picture shows cumulus over the Irish Sea in November. The convection may continue day and night and with so many thermals being evaporated into it the air becomes damper, and fragments of cumulus tops which have spread out *(See* 7) can be seen. The smoothness of these layers of cloud, called *strato-cumulus*, shows that the "boiling up" motion of thermals is not taking place within them and the air is therefore said to be *stably stratified*, or *stable* for short.

4. A *Pileus*, or cap cloud, often appears on the top of a thermal. The air above the thermal is pushed up as the thermal arrives and if it is near to its condensation level a cloud may be formed in it. We can see in this picture that there is a stable layer because the haze has a fairly sharp top which the convection of the previous day did not penetrate. The air just below this stable layer, or *inversion* as it is often called, is nearly saturated, so that a little lifting produces cloud in it. The cloud cap is smooth because the air is stable.

This cumulus is being produced from thermals rising off the east side of a hill which is the first place to be warmed in the morning. It is drifting from there towards the east in a light wind and is being followed by more thermals from the same source which have not risen so high yet.

5. *Artifical thermals* may be produced by a sufficiently large artificial source of heat. Here a group of three power stations close together, near Coventry, are sending up thermals and forming small cumulus clouds, the shadows of which can be seen on the dusty air below.

This is a very dry day, in an anticyclone, or high pressure area, in which the air is slowly sinking. Just as the air eventually reaches its condensation level if it rises, it is made drier by sinking, and the thermals seen here do not produce large cumulus. Undoubtedly the moisture from the cooling towers makes the condensation level a little lower than in the thermals rising from the sun-warmed ground around.

High up in the sky is the remains of an artificial cloud of ice particles formed in the exhaust of an aircraft *(See* 36). Ice clouds evaporate more slowly than clouds of water droplets and therefore may persist for much longer without any renewal taking place *(See* 19), and this trail has been spread by the air movements over a large area of sky.

6. *Cloud streets* is the name given to parallel almost equally spaced lines of cumulus cloud. They tend to lie more or less along the wind and are most common over land when the convection has reached up to a stable layer in the morning, but has not yet penetrated it. Thus the convection is taking place in a layer of air of more or less uniform depth.

7. *Anvils* are formed when cumulus reaches a stable layer and spreads out horizontally at it, if it is not warm enough to penetrate it. In this picture we are looking at the under side of the overhanging part of the cloud where it is spreading out. It is composed of water droplets, and is sometimes called "strato-cumulus formed by the spreading out of cumulus" but the simple name *water anvil* is better. After the original cumulus has evaporated, these flattened portions often remain for some time longer because the motion in them—and consequently also the mixing with the surrounding drier air—is slower than in the cumulus *(See* 3).

The underneath side of a water anvil, where cloud lies above unsaturated air, often becomes wrinkled and threatening (like the mamma shown in 15) but this does not indicate that there is rain anywhere.

8. An *inversion*, or stable layer, becomes clearly visible on a calm morning when the warm dusty gas from a large chimney rises up to it but is not warm enough to penetrate it. Any slight wind then carries the plume off to one side at the level of the inversion.

9. In an *anticyclone* (as we saw in 4) the air is sinking (very slowly). While the higher layers are sinking, the lowest 3000 ft or so may not be. In this picture we see the cumulus cloud formed over the sea (mid-Atlantic), and the spreading out level under the sinking layer *(See* 7) is so near to the condensation level that the thermals are not visible as cloud until the flattening out process has begun. The cumulus are then rather like pancakes in shape.

If the convection continues for long enough it will fill the whole sky with these pancakes, which can then be called *strato-cumulus*, i.e. a layer of cumulus.

10. *Warm rain*, seen here as dark shafts between the clouds and the horizon, is common in cumulus *over the sea*. If raindrops are to be formed, each drop from the moisture of many thousands of cloud droplets, they must remain inside the cloud for the best part of an hour at least. The upcurrents in land cumulus often carry them to the top in a few minutes and they are then evaporated, but maritime thermals are less vigorous and so light rain showers may be formed.

The rain is called warm rain because no part of the cloud is frozen. This type of rain is commonest in warm climates.

11. *Warm rain* may also occur *over land*. In this picture we can see that the sun is low (the rainbow arc is nearly vertical) and the convection feeble. The exterior of the left part of the cloud is blurred by the falling raindrops while the right half is sharper and, in fact, has no rain falling from it (it produced no rainbow when the wind carried it to the required position).

12. *Glaciation* occurs when the cloud is sufficiently cold. The higher thermals ascend, the cooler they become. The droplets do not freeze until they are much colder than 0°C, but remain as *supercooled water* for some time. When some of the drops freeze, perhaps because they are the larger ones, and the ice crystals fall out, they do not evaporate as quickly as the part of the cloud composed of water droplets. The small thermal on the left in this picture has trails of ice falling from it. They are much whiter than trails of water droplets and could show no rainbow.

The large tower on the right has pileus on it.

13. *Cumulonimbus* is the name given to cumulus from which rain is falling. The clouds in 10 and 11 are therefore cumulonimbus. This picture shows a much larger example (the same cloud as in 12) which is partly glaciated at the top. The rain can be clearly seen.

At a much higher level a thin layer of *cirrus* cloud, which is cloud composed of tiny ice crystals, illustrates how the resistance of the ice crystals to evaporation causes the cloud to assume a fibrous appearance. The fibres are partly due to the fall of the particles, as in 12, and partly due to the stretching out of the cloud by the air motion.

14. *Cumulonimbus* produces an *ice anvil* if it spreads out at a stable layer after becoming glaciated. The stable layer in this case is the *tropopause*, or base of the stratosphere (the permanently stable, i.e. stratified, layer of the atmosphere). The nearest tower has pileus on it as it pushes up a short distance into the stable layer before sinking back and spreading out as an ice anvil. Ice anvils usually show the characteristic anvil shape better than water anvils, and do not evaporate as quickly, particularly at their thin edges.

15. *Mamma*, the white protuberances in the lower centre of the picture, are formed on the under side of the anvil of a cumulonimbus if there is a substantial fall of ice particles out of it. This picture shows the edge of an anvil illuminated obliquely by the setting sun. The falling particles have been carried up in the upcurrent but are no longer supported when it spreads out horizontally.

The up-motion in the cumulonimbus sometimes produces a temporary down-motion in the surrounding air which causes evaporation of cloud, and here we see part of a clear ring which surrounds the anvil as it pushes out sideways in an already existing layer of cloud.

16. *Castellanus* clouds are cumulus which are not formed directly by thermals rising from the land or sea, but are produced mainly by the heat of condensation when cloud is formed. They look smaller than ordinary cumulus because their starting point is at the cloud base, whereas ordinary thermals have grown considerably in rising from the ground. They usually look taller than cumulus, and are often in groups of several together rising from a thin line along which condensation is occurring. Much of the cumulus over the oceans is castellanus (e.g. 10) because the upcurrents from the lukewarm sea are much feebler than those produced when condensation of cloud begins.

17. *Castellanus* may occupy a large part of the sky if upward motion and condensation is being produced over a wide area by converging horizontal winds or by the passage of the air over a hill. These clearly defined bases are typical of castellanus, which, as in 16, produces thermals much smaller compared with their height than ordinary cumulus.

18. *Fallout* of ice crystals often takes place from castellanus which are cold enough to become glaciated. The streaks of ice crystals are clearly distinguishable by their fibres and transparent silky whiteness from the sharp-edged but ragged clouds of supercooled water droplets from which they are falling.

The ice cloud is thus at a lower level than the water cloud in this case.

The Clouds of a Low Pressure Area (Depression)

19. *Cirrus* formed at higher levels, i.e. at sufficiently low temperatures for the water droplets to freeze as soon as they are formed shows the fibrous form without any typical water cloud being present at all. Although these clouds look windswept they change their shape very slowly by comparison with water clouds.

Below is a *wave* cloud—a cloud formed in the crest of the wavy motion produced when the air flows over a hill. This is a water cloud, and is in the shadow of some cirrus.

In the distance is a more continuous layer of fibrous ice cloud which is therefore called *cirrostratus*. There are some small fragments of cumulus, but the shadow of the cirrus is cutting off the sunshine so that there are scarcely any thermals to renew the evaporating cumulus. There is great contrast of colour here because the picture was taken using a polarising screen which, in certain directions, excludes the blue sky light.

20. As a *low pressure area (depression)* approaches, first we see cirrus perhaps like 19, and this gradually thickens into a complete layer of *cirrostratus* as shown here. Haloes may be seen in these ice crystal clouds; the most common halo such as is shown here subtends an angular radius of 22° round the sun (or moon).

Below are some fragments of middle level cloud, which are often produced in lines parallel to the line of the advancing *warm front*.

21. As the warm front of a depression approaches the cirrostratus is often supplemented by layers of water droplet clouds called *altostratus* (high layer). The water droplets blur the halo and the sun is darkened. *Wave clouds* are formed over hills in the lower layers (because the wind and temperature structure of a warm front are very suitable for the formation of waves) and remain more or less stationary over or near to the hill producing them, with the air blowing in at one side and out at the other (*see* 28). When the rain begins altostratus is called *nimbostratus.*

22. *Stratus* is often encountered when the warm air, which gives the warm front its name, at last arrives. It is usually air of tropical origin which has a very low condensation level when it has been cooled by travel to more temperate latitudes. Often there is drizzle or light warm rain on hills in the warm air, and if the cloud is in contact with the hills it is called *hill fog.*

23. *Warm sectors* of depressions (low pressure areas), i.e. where the warm air is, are not always drizzly like 22, and inland in summer the sun's heat penetrates through the thin layer of low cloud; by warming the ground, it starts up thermals which gradually evaporate the stratus and produce cumulus, as in this picture. When the cumulus is combined with the cirrus of the depression some of the most beautiful skies to be seen in temperate latitudes are displayed.

24. The cold air returns when the *cold front* arrives. This is often a *line of cumulonimbus* formed where the cool air lifts the warm damp air. Although not all cold fronts are accompanied by the kind of refreshing clearance of storm clouds shown here, they are the most interesting and produce very rapid, beautiful, and dramatic changes in the sky. In this example mamma can be seen in the anvil of the nearest cumulonimbus. The cold air is moving from the right.

25. *The clearance at a cold front* is often quite sharp but not accompanied by a belt of rain. On this occasion the low cloud disappeared first and in this picture we see the final edge of the cirrus which is moving away to the left. In the cold air on the right cumulus is rapidly developing as soon as the sunshine begins to warm the ground. This picture was taken at about 11 o'clock on a summer morning; and further from the front where the cold air was deeper, showers occurred in the afternoon.

26. *Cumulonimbus in deep cold air* is here illustrated with the lower clouds leaning forward (to the left) showing that the wind is stronger at higher levels. The anvil has snow and rain falling from it, and this gives it the softened outline, while a newly growing part of the cloud penetrates, with sharp white outline, far above the spreading anvil. At the top of the picture we see the evaporating edge of smaller cumulus.

This cumulonimbus is typical of cold windy days in *polar air* that has recently come from a cooler latitude. By comparison 14 is a typical cumulonimbus of the tropics, and reaches to a much greater height.

Wave Clouds, Billows

27. *Clouds at three distinct levels* are seen here. The *cumulus* is typical of hilly country—it forms predominantly over the higher ground (where the heat of the sunshine is put into the atmosphere at a higher level), and is tending to form streets along the wind. Across the middle of the picture is a *wave cloud*, which has a smooth outline, and remains almost fixed in position in the sky as the wind blows through it. It is a water cloud at about two miles above the ground. At a height of about five miles are long streaks of *cirrus* along the direction of the wind.

28. *Wave clouds* are common at sunrise but often disappear on sunny days at inland places, and return again in the evening. They exist where the air rises above its condensation level as a result of passing over a hill. The hills are fixed and so the pattern of flow may also remain fixed with the crests of the waves stationary. These wave clouds are at the top of a hazy layer similar to that seen in 4 where a pileus cloud was formed by a thermal, and had the same smooth appearance as a wave cloud. The wave crests are faintly visible, even where no cloud is formed, when they are viewed obliquely: note the continuation of the wave crest between the two wave clouds in middle distance.

The nearest cloud is composed partly of *billows* which are formed at the inversion where the air above is sliding over the air below, and the billows are in the position of rollers between these layers. They move along with the wind so that new billows are formed at the near edge of the cloud, pass through it, and evaporate at the lee edge. Billows can be distinguished from waves by the fact that they move with the wind, whereas if a small piece of cloud is watched carefully it can be seen to move through a wave, appearing at the upwind edge and evaporating when it arrives at the lee edge.

29. A *wave hole* is produced in a layer of cloud where the air descends into the trough between two waves.

This picture also shows that wave clouds, with their typical smooth appearance, can occur among cumulus clouds. This is particularly true near the coast, because as the air blows inland it takes some time for the cumulus to develop. Further inland the convection tends to spoil the smooth flow needed to produce wave clouds.

30. A *plume of cirrus* is often formed in a wave cloud if the temperature is sufficiently low—that is if the cloud is at sufficient height. In this picture the smooth upwind edge of the wave cloud can be seen, but the water droplets in it have frozen and the ice crystals do not evaporate; they stream away as a long plume in the wind. This wave cloud is at a height of about five miles; but the low wave clouds, which are dark because the sun has set behind the mountains, are at about 3000 ft. They only appeared when the sun went down and the convection ceased.

31. *Streets* and *lee waves* are not usually seen together; but in this air view we see streets (similar to 6) on the right, with the wind blowing along them from the right. The coast lies almost across the middle of the picture, and the air descends to the sea over some fairly high cliffs: this sets it into oscillations so that it executes a series of waves as it moves over the flat sea. Eight or nine lee wave clouds, which are in the crests of these waves, and parallel to the coast, can be seen in the distant part of this picture, particularly on the left.

32. *Billows* look rather like waves, but if watched carefully are seen to be moving along at about the same speed as the wind. These are formed in a thin layer of cloud and may be termed *altocumulus* because they are due to convection in the layer which is far from the ground. Once the convection has broken the layer into cloudlets very little further change in form takes place.

33. *Billows* occurring on the tops of cloud layers often go through a development which looks like water waves breaking on the beach. The lowest clouds here are produced in the air rising up the mountain face as it moves inland, and on the top of it the sliding of the air above over the cloud air causes the rolling motion of billows to develop and the complete cycle of the "breaking" of these little billows and their final disappearance again into the cloud surface takes place in about a mile in a period of two or three minutes. We see here a cross section view of the kind of billow motion seen from below in 28.

Ice Clouds

34. *Cirrus* may take a variety of forms. On the left we see *feather-* or *leaf-like* cirrus, with the ice crystals sinking rather rapidly along the spine but slowly along the edges. On the right we have *lines* of cirrus (similar to those in 27) along the wind which was towards the top right of the picture. Across these, and at a lower level (which could be seen because they were darkened earlier at sunset) are *striations* across the direction of the wind which are produced in the same way as 35.

35. *Hooked cirrus* is formed when ice crystals fall from a source cloud, which often looks like a small piece of castellanus (*See* 18). This *fallstreak* is drawn out over four or five miles because the wind at different levels is different. The particles are falling at 2 or 3 ft per second.

36. *Contrails* (short for *aircraft condensation trails*) persist when they are formed at a sufficiently low temperature for them to become frozen before they evaporate. Then (as in 5 and 35) because of different winds at different levels, they may be spread over the sky to form a layer of *artificial cirrus.* Sometimes the air is sufficiently dry for the ice cloud to evaporate in ten minutes or so; but often it may last for several hours.

Below, dust rising from a power station can be seen; as the sun goes down the natural convection from the ground stops so that the solitary *artificial thermals* can easily be seen. They can also be seen well early in the morning.

Coloured Arcs

37. *Mock suns* are often seen as bright patches of light, with the edge towards the sun tinged orange, at the same elevation as the sun and just outside the common halo (20). This one, seen through a telephoto lens, is formed in a contrail. Mock suns only appear in ice clouds.

The brightest and most colourful mock suns are seen at sunset from an aircraft flying at the same altitude as the ice cloud which forms them (or in polar regions where the ice cloud is close to the ground).

38. *Circumzenithal arcs* are often seen in ice clouds when the sun is low. They are not usually long lasting, and because they are rather high in the sky they are often not noticed. The centre of the arc is vertically above the observer, and the colours are very bright with red towards the sun. They may be seen in many different forms of cirrus including contrails and the thin edges of ice anvils. There is a contrail in this layer of natural cirrus.

39. The *glory* which surrounds the shadow of the observer when it falls on a cloud can only be formed by water droplets, not ice crystals. It becomes rather faint if the cloud is vigorous cumulus and is best developed in fog or patches of cloud formed on hillsides up which the wind is blowing. When a glory or a rainbow (11) is seen we know we are looking at water droplets and not ice crystals. Similar colours round the sun or moon (called a *corona*) also indicate water cloud, as do the coloured patches of cloud somewhat further from the sun (called *iridescence*). Glories are very easy to see from an aircraft flying above cloud. The corona and iridescence are best seen through dark glasses.

Cloud Layers

40. Layers of *altocumulus* often have very sharp edges whose explanation is rather uncertain. Each of these layers is composed of cloudlets because thin uniform layers cannot exist for more than a few minutes without breaking up in this way. Convection in them is often caused by the loss of heat from the top of the cloud into space. Sunshine has very little effect: most of the energy of sunshine is either reflected or passes through the cloud.

If there is a difference in wind between the top and the bottom of the layer the cloudlets are formed into rolls (*See* 32).

41. *Stratus* and *Altocumulus* layers at dawn show that the air is stable. The wind is light so that there are no waves visible in the altocumulus layers above (waves would form shapes like 28 and 29 if they were present). The layer of *stratus* below is at the top of cool air which fills the valleys and which does not move with the wind above; little fragments of cloud are being carried off the top by the wind.

As the morning develops the warming of the ground by sunshine produces convection in the valleys: the air there becomes warmer and the stratus is evaporated. Later, as the heating proceeds, thermals become warm enough to rise through the inversion which lay over the stratus at dawn and cumulus is then formed over the mountains.

42. *Radiation fog* is the name given to stratus which forms over flat ground or in valley bottoms at night. The cooling which produces the cloud is the result of a loss of heat by radiation into space from the ground, with no compensating heating from the sun. Only a very shallow layer of air close to the ground is cooled and so the cloud is very shallow. It is soon evaporated by sunshine in summer but may last all day in the depth of winter.

To cool a deeper layer than this one it is necessary to cool it at the top so that downward thermals (cold ones) are produced, or for the cooling to be spread upwards by srirring motions in the air.

43. *Sea fog* is common in the warm sectors of depressions. It is also common when tropical air travels over cool sea in other circumstances. Because the sunshine has a negligible heating effect on the cloud itself (*See* 40), and alters the sea temperature very little (except where the water is very shallow), sea fog is only dispersed when it is carried in the wind over sunwarmed land. Here we see the fog over the water of an estuary being blown on to the land and gradually turned into cumulus. Above is a layer of stratocumulus probably formed by thermals spreading out at that level.

Wet Grass

44. *Wet spiders' webs* are found when they have been inside a cloud. The tiny fibres catch the cloud droplets whereas larger objects deflect the airflow with the droplets in it and do not capture them. The droplets then amalgamate on the threads into larger drops. This is not dew deposited, though a web might act as a centre for the condensation of dew if it were in the air cooled within an inch or two of the ground.

45. *Guttation*, which is the exuding of water from the tips of the blades of grass, occurs on summer nights when the ground and the roots of the grass are warmer than the air. This is not dew, and can easily be distinguished from it because dew is deposited on grass in the form of very many tiny droplets along the edges of the top few inches of the blades of grass. A good example of *dew* is often to be seen on a car roof, and the moisture comes from the air.

Violent Storms

46. A *tornado* is the most violent kind of storm that can occur in the atmosphere. The wind rotates round a centre of low pressure which is formed under some of the most rapidly growing cumulonimbus clouds. The surrounding air is sucked into the vortex, as is clearly shown by the smoke in this picture. If solid objects are sucked up or torn from the ground the violent rotation causes them to be thrown outwards in all directions.

Because of the very low pressure in the centre of the vortex cloud is formed there, and if it passes over a house the excess of pressure inside causes the walls to explode outwards.

47. A *hurricane* may not be as violent as the vortex of a tornado but it covers a circular area a few hundred miles wide. This picture taken on 13 September 1961 from a Project Mercury capsule from a height of about 100 miles shows the west side of a hurricane over the Atlantic ocean. In the centre of these rotating storms is a fairly clear calm region called the *eye*, where the dry air of the stratosphere is sucked downwards. The eye of this hurricane can be seen on the right of the picture.

Placid Scene

48. *Strato-cumulus* is here seen over the ocean beneath an inversion with *cumulus* at a higher level over the mountains inland. The Mercury capsule was at this moment a little more than 100 miles up looking NE'wards along the coast of Africa towards Agadir from just east of the Canary Islands.

The wind is from the north, and the warmer sea in the lower half of the picture forms cumulus which spreads out (rather like 9), below an inversion and is in *streets* along the wind. Where the cloud comes over land it is evaporated by the convection, so that the line of the coast is clearly visible. The cumulus over the mountains is beyond (to the NE of) Agadir.

The coast can be traced beyond Casablanca to Tangier and the coast of Spain and Portugal is faintly visible on the horizon.

PRACTICAL STUDIES

THESE notes are not intended to be complete, but rather to be a basis from which further work can be developed.

1. The area of sky visible from the ground is very large. The horizon seen from a height h is at a distance roughly equal to $\sqrt{(2Rh)}$, where R is the earth's radius. (Note: R and h must be measured in the same units in this formula. If, however, h is given in feet the distance is about $1{\cdot}2\sqrt{h}$ miles, or if h is given in metres the distance is equal to about $3{\cdot}6\sqrt{h}$ km.)

When seen from a height h a cloud at a height H disappears below the horizon on the sea when it is beyond a distance of about $\sqrt{(2R)}(\sqrt{h} + \sqrt{H})$.

Thus, cirrus can still be seen above the horizon at a distance of 150 miles if the visibility is good enough, and from a good vantage point the clouds over most of "a county" can be seen.

2. The speeds of movement of clouds, particularly small cumulus, can be measured by watching the rates of progress of their shadows across the countryside. Distances can be estimated from a large-scale map or by some other means.

3. The positions of clouds can be estimated from the positions of their shadows or inferred from the positions of hills which may be causing them, and their heights may then be estimated from the angle of elevation by simple trigonometry. If the height of cumulus base is known (by comparison with a mountain of known height, for example) and appears fairly uniform, the height of the tops may be estimated from the angle of elevation. Thus, approximately, if the base is 5° above the horizon and the top of the same cloud is 15°, it is at three times the height of the base.

4. Angles may be measured crudely by getting to know what angle the extended hand subtends, or simple protractors and plumblines can be constructed. Sextants and theodolites may also be used if available, but much can be done without them, and the very simplest trigonometric calculation can be very informative.

5. If the speed is known (e.g. by method 2) the height can be estimated from the apparent angular motion when observed from a fixed point. If the height is known the speed may likewise be computed.

6. If the distance is known approximately the rate of rise of cumulus tops can be measured by observing the rate of increase of elevation of the top. This may be done by watching the cloud top move up beyond a pole (with lengths marked on it) at a known distance.

7. The wind direction may be observed by watching the movement of the clouds across the top of a pole while standing at the base.

8. The highest clouds remain illuminated longest and are whitest at sunset, and clouds nearer the sun cast shadows on those farther away. The sun shines upwards on them at sunrise and sunset and so these are the best times to observe cloud bases, and mamma in particular.

9. There are many interesting questions which may be answered by these simple methods of observation e.g. (i) Do lines of cirrus always lie along the wind direction? (ii) What happens to pileus cloud—it only lasts for a few minutes? (iii) Does the wind blow exactly along cloud streets? (iv) Is the motion at a sharp cloud edge (25, 40) along it or across it? (v) Is there any downward motion on the outside of a cumulus cloud? (vi) If the clouds at different levels are moving differently is this related to the orientation of billows and fallstreaks? Are billows usually at right angles to fallstreaks?

(Allowance must be made for perspective if they are not vertically overhead. Look at right angles on a ceiling to get an idea of how they would look in the sky.)

10. To observe the growth and development of clouds, particularly cumulus and cirrus, the following methods may be used:

(i) sketches, using pencil and paper, at intervals of 5 or 10 minutes.

(ii) photographs taken a minute or two apart.

(iii) single exposures at a regular interval of 2 or 3 seconds on cine-film with the camera fixed, shown at normal cine-projection speed. This can be particularly revealing when applied to wave clouds.

Times and places should always be noted at the time of observation because it is difficult to remember details when several occasions have been observed.

Pictures may be used as a basis for trigonometrical calculations if known landmarks are included and (in the case of photographs) the focal length of the lens is known. For instance the angle subtended by haloes, and other coloured arcs can easily be measured from photographs.

11. Use sunglasses to examine the clouds near the sun, where many interesting colours and shapes can be seen.

It is of little value to use binoculars except on such tiny clouds as contrails. Do not risk looking at the sun through them except when it is *very* low and red, but then it can be very interesting. When the sun is not low it is very dangerous for the eyes.

12. The sky changes most rapidly at sunrise and so in an hour's observing then you will usually see more changes than at any other time. Sunset is the next most interesting time. The changing light keeps revealing new feature of the clouds; the changing heat causes developments of cloud forms.

13. Use a polarising screen, or observe the sky by reflection in a piece of plate glass (preferably black glass), to reduce the intensity of the blue of the sky and increase the sharpness of the clouds. The movement of clouds may be sketched by means of wax pencil on a horizontal piece of glass observed from a fixed point and computed by simple trigonometry afterwards.

14. Red or orange filters make the clouds show up against the blue sky better on monochrome film. No colour filters should be used with colour film, although a polarising screen is often useful.

15. Record details of photographs at the time so that repetition of mistakes can be avoided. Do not begin by trying several different makes of film, but learn by experience to get the best possible results from one or two makes: good film can produce poor results by incorrect exposure.

16. Do not hesitate to take pictures towards the sun if the sky looks interesting. (If your camera has a focal plane shutter which is not protected by a mirror do not point the camera at the sun with the lens open for more than a fraction of a second or the shutter blind may be scorched.) Reduce the exposure considerably for haloes, etc. Exposures of a few seconds on colour film after sunset can be very interesting.

17. If wave clouds are really stationary, would not a time exposure of a few minutes be possible by moonlight?